A Five Years' Whaling Voyage 1848-1853

A FIVE YEARS' WHALING VOYAGE

1848-1853

J. C. Mullett

Ye Galleon Press
Fairfield, Washington
1977

Library of Congress Cataloging in Publication Data

Mullett, J. C.   b. 1830.
 A five years' whaling voyage, 1848-1853.

 Originally published in 1859 by Fairbanks, Benedict, Cleveland.
 1. Whaling. 2. Seafaring life. 3. Voyages and travels. I. Title.
G545.M94   1977   910'.41   (B)   77-20268
ISBN   0-87770-182-2

This old woodcut shows the deck of a Pacific whaler, early nineteeth century. Note small size of the ship, furnaces for rendering whale oil and wooden barrels used for storage.

PREFACE

This narrative is not written for the purpose of deceiving the community, or to impress upon the minds of the public any untruths; but rather, a true relation of the incidents and hardships experienced by the Author in his travels over the trackless main, from the Arctic Regions to the Southern Ocean, visiting almost all the islands of the Pacific, and various other parts of the globe, and during this time have become acquainted with many of the habits and customs of different nations, both civilized and uncivilized, with those of the savage islands of the Pacific.

J. C. Mullett.

Decades ago, sailing ships of modest size searched the oceans for whales. Whale oil lighted the houses and streets of this country before petroleum oils were used in their smokey lamps. Long before the days of radio, radar and loran, small vessels dependent upon the vagaries of wind and weather penetrated the Arctic and Antarctic Oceans in search of whales. In those long ago days men hunted whales from small wooden boats. The men used wooden lances tipped with iron points. Deep, cold ocean became the grave of many a brave whaleman while hunting these great mammals far from home.

The J. C. Mullett book is No. M-886 in Wright Howes U.S.-IANA.

was born in England, in the town of Weymouth, county of Dorsetshire, in 1830, and lived there until I was eighteen. My father was a tradesman, one of what is called middling class in society. I attended school until I was twelve years of age. At that time my father died; my eldest brother then took charge of the business, and I commenced to learn the trade. He remained three years in charge of the shop, when he bade me and my mother adieu, and took his departure for the United States, leaving us alone. The business then fell upon me. After the lapse of two years I received a letter from my brother in the States, advising me to leave my native land and join him in America. My mother did not like the idea of being left alone, but after considering the matter over, and some persuasion, she gave her consent. At the age of eighteen I took my leave, and went to the city of London. I there engaged passage on board the bark *Orleans*, of St. Johns, bound for New York. There were about three hundred and fifty passengers on board. We had a pleasant passage, and in twenty-six days after our departure from the British Channel the anchors were cast off Staten Island. I did not go on shore until the next day. I then started for the city, where on my arrival I fell into conversation with a person, who, after learning what country I was from and the place of my destination, informed me that he had joined a whale ship, and on the morrow he was to sail. He also persuaded me to accompany him, and to induce me to go assured me in three years that my part would not be less than a thousand dollars. This soft tongued, black hearted schemer plead with me until I consented to go with him. Without delay we went to the shipping office; my name was signed on the ship's articles, and I was entrapped before I was sensible of the act. It being near night I and my friend returned to a boarding house, where we tarried until morning, when we breakfasted together, and about nine o'clock I found myself on board the old steamboat, making her ways towards New Bedford. I then looked around for my friend, but he was nowhere to be found. I began to mistrust that I had been gulled and saw that all his sophistry and friendship was nothing but a bait used to entice me into the trap by which I was caught. I found myself with three or four others in charge of another man, and on our arrival at New Bedford we were commanded to put up at Carter's boarding house. I did not see much of the place until the next day, and then I saw nothing but oil casks. I visited the wharfs and at last found the ship in which I had enlisted. She was called the *George and Susan*, and was commanded by Captain White. I slept on board, and I doubt not felt as big as Commodore Perry did when he gained his victory on Lake Erie, but the time to test by bravery had not yet arrived. The sixth or seventh day after our arrival at New Bedford the sails were spread and the old George might have been seen looming her way through the waves of the mighty ocean. There were thirty-three hands on board, who all appeared to be much pleased in their present situation, but their time was coming as well as mine. There were eighteen on board who had never before trod the deck, and as the ship began to roll with the waves, they were sensible that the legs that carried them about on land were not sea legs. I sympathised with these men, though I was not much in advance, but I was brought up by the water and my voyage across the Atlantic had made me quite a mariner, and I felt as much above them as Bonaparte did over a marine corporal. That night the watches were chosen, and I found myself in the captain's watch. I had the midnight watch, and stood my first trick at the wheel, and there I began to view the situation. I thought of home and my friends, and my heart sank within me. My meditation drawed my attention from the compass, but in spite of all my efforts the ship run from her course, and I was aroused by the captain of the deck, whose voice came like a clap of thunder with the words, "Mind your wheel there." They were new to me, and coming so sharp, that it took me completely back, and the ship was soon in the same fix, for she had swung around, and her head was pointed towards New Bedford. She was soon brought on her course again, and no damage ensued, but I was very much frightened and received a severe reprimand from the officer. Eight bells rung, the watch was called and I of course was at liberty. I retired to my berth, but not to sleep, for soon the winds began to blow and there being but a few men who were able to do duty, the watch below was again called. This was also new to me, and I must say that I was unwilling to go, but the orders had to be obeyed;

we therefore came on deck, and such a scene I never beheld. The sea was washing over the bulwarks, the fore top gallant mast was already gone by the board, the waves rolled mountains high and the stroke of a single surge made every timber tremble, and caused the vessel to quiver like an aspen leaf. How frail an object is the stoutest ship when in the fatal grasp of an ocean tempest. With what speed it is driven before the resistless force of the wind. The wild foaming surges, the lurid lightning, the crashing thunder, the rolling of the ship, the strained and creaking ropes, the bending masts, falling spars, rent and torn sails, the cold mists that fill and darken the air, the consternation of rapidly beating hearts, the dread and horrible suspense of the hour filled my heart with terror. The remainder of the canvas was taken in, the companionway was fastened down in order to keep those that were sick from the storm; four of the stoutest men on board were lashed to the bulwarks. I enquired of the mate if he knew where we were. He informed me that we were in the Gulf Stream. The gale was favorable for us and the ship kept on her course. We expected every moment she would be swamped. We remained in this situation for twenty-four hours when the storm abated.

The sea was smooth and we once more glided towards our destined port. Those that were seasick began to recover and were once more permitted to promenade the deck, and the men were busily engaged in repairing the damage. We spoke the bark *Henry Clay*, from Fair Haven; she was three days from port and had not escaped the tempest which we encountered; her light spars were all carried away, but no lives lost. The captain came on board and got from us some spare masts and yards. After passing the usual compliments we bade them farewell, and made our way onward. Preparations were now made for taking whales; the lookouts were sent to the mast head. This I thought useless, more especially when it came my turn, for I did not know the spout of a whale from the spray of a porpoise, but I soon saw the distinction between the two. But perhaps before I go further it will be well to inform my readers what the implements are which are used in capturing whales. There are generally four boats, each boat taking six men, the officer, the boat steerer, and four foremast hands. Generally there are six or eight harpoons, three lances, and one boat spade, all of which have sockets in which a pole is attached. These poles are from five to six feet in length, and the harpoon about three, and when the boat line is attached it is used for holding the boat to the whale — the lances are used for killing. This was the work for all hands at this time, to prepare the boats for action. The boats' crews were chosen and soon were receiving instructions from their officers in regard to rowing and so forth. The first calm day we had, the boys received a training. An empty barrel was thrown overboard, the main yard was hauled aback, the boats lowered away, and pointed towards the barrel, which was about a mile astern. The officers urged us on by saying "spring, boys," "pull, my good fellows," "we gain on him," "we shall strike." The boat steerer was told to stand up— "give it to him" cried the officer. "Stern, stern," was the cry, "we are fast." Every man backed water for his life.

I had been so excited by the conversation of the mate that I entirely forgot myself. I imagined I was fast to one of the largest whales that ever traversed the sea, and my heart stood on end with fright, but I chanced to get a sight of the barrel, when my pulse ceased to beat so rapidly and I became more calm. The officer asked us how we liked our first experience. We told him that if we always came off as safe as that we should like it first rate. He told us that there was no danger, and we then returned to the ship. On reaching the deck I felt myself at home for I had got to be quite a mariner and felt myself one of the most experienced in the whale fishery. The boats were hoisted and the ship put on her course.

On the twelfth day of our voyage we raised a school of sperm whales, the boats were lowered away and we chased them about three hours, but without success, and the boats were turned to the ship. I for one was heartily glad, for there had not been but a few works spoken through the chase. I found myself very courageous when going towards the ship, but I will leave you to imagine how I felt when going towards the monster. We reached the ship, but the captain said if we had not been so scared we might have caught two or three of them, but now was the time for me to play my part. I stood erect and appeared as bold as a lion, but I felt as meek as Moses.

Not many days after this on one of those bright moonlight evenings when the sea reflects every object upon its surface, I stood leaning upon the taffrail, gazing at the sparkling waters; my eye turning to the officer, I enquired if he could tell me what that was. A shark, he replied. There was the monster that I had so often wished to see. He appeared very composed and kept up his row with perfect ease. My curiosity and attention were aroused and I wondered if there was any way of capturing him. At length

I informed the mate of our visitor, who told me to catch him, but I replied that I had no hook. After calling me a greenhorn and laughing at my speech, he informed me how the visitor might be brought on board. I then fastened a tin cup to the end of a string, calling all the necessary help, then dropped the cup into the water, where the reflection of it soon caught his eye. Bent on his schemes he made for the cup, but instead of getting what he expected, the harpoon was sent through him and clinched on the other side; three or four heavy pulls by the hands and he was landed on the deck.

I presume to say that most of my readers have made themselves acquainted with the passage in the good book, "do good unto your enemies," but I saw no one there who I thought manifested a Christian disposition, every man with a club or billet of wood, taking satisfaction out of John shark's hide, until he departed this life. I once heard a story of a marine soldier falling from his station, and never afterwards was seen. A short time after this the sailors caught a shark; on opening him they found the marine's coat and cap undamaged. Not knowing but this fellow had caught the marine, my knife was applied, but I found nothing to my satisfaction. He had four rows of teeth and I never saw anything better calculated to grind up human nature in my life. I then took his length which was ten feet. At this the watch on the deck was relieved and the shark was thrown into David's locker. All this watch my joy had been full. I returned to my berth and I doubt not but I felt as much of a conqueror as General Jackson after the battle of New Orleans.

No incident worthy of note occurred for two or three of the following weeks, during which time the weather was favorable for us, and we were rapidly sailing towards Cape Horn, a place I much dreaded from the stories I had heard about it. The weather was growing cooler and the winds stronger. I enquired of the mate where we were, he said we were nearing the cape. The men were called down from the mast head and all things were made ready in case of a storm. I well remember the morning that we doubled the Horn, my watch had been eight hours on deck the previous night and was about getting my breakfast and enjoying a good forenoon's sleep when the captain, a large six foot, black eyed down easter, made his appearance on deck, and on going the usual rounds looking at the compass and enquiring what the weather had been through the night, &c., he commenced promenading the quarter deck, scanning the horizon, when his eye caught an object which proved to be a sperm whale. "Back the main yard," he exclaimed, "and call all hands." The boats were lowered away and the chase commenced. It was a daring piece of business; even the officers trembled at this command, but he was the captain, and his orders must be obeyed. We chased the monster about one hour and should have struck him in a short time, had it not been for a storm which had been gathering in the distant horizon and was now close upon us. The colors were set for us at the mast head for our return to the ship; every heart leapt with joy as they saw the flag flying at the mast head. It was the second time we had given chase without success; our backs were turned towards the monster and we were making our way towards the ship as fast as possible, but being some three or four miles distant the squall struck the ship before we could reach her, we being about one mile and a half to the leeward at the time. The men being but few on board they were not able to manage her; we saw the sails rent to ribbons, and expected that the George and Susan would be crushed to pieces. A man was thrown from the jib-boom while in the act of securing the flying job; the life buoy was thrown to him which he grasped with eagerness; turning his back to the storm the waves carried him swiftly away. We saw from the boats the whole transaction without rendering any assistance, for the storm had already met out little skiffs, the boat crews were commanded to lie down, the officer saying if he could keep her head towards the storm the boat would live through it. We obeyed his orders, and in a short time the squall passed by; as the colors on board the ship were at half-mast as the signal of distress, we wasted no time but made all haste to their assistance. After rowing a short time we saw the man who had fallen from the boom still clinging to the life buoy—every muscle was strained to its work to save the perishing mariner, we at length reached him, he was benumbed and unable to speak. On reaching the ship he was taken to the cabin and after the lapse of three days he was able to walk the deck. The captain was very much frightened, and said he had never experienced such a squall without doing more damage, and declared he'd never lower his boats again off Cape Horn. The torn canvas was taken down and new sails put up in their places, and every hand was busily engaged in repairing the damage. There had been a ship seen to the windward of us before the storm; our captain said he saw her in his telescope before the squall struck her, she went out of sight in an instant, and he thought she was dismasted or completely wrecked, as we never saw her again. But after the repairs were finished and we were once more floating quietly upon the bosom of the sea, I had time for reflection, and thinking over past scenes and the danger we had been exposed to, the day previous, I became satisfied that I was not calculated for the whaling business, and I came to the conclusion, that the first port we made, was it inhabited by savages or civilized people, to make my escape, but one thing was certain, there was no escape now, therefore I resolved to make the best of my situation, and keep my resolution to myself. After doubling the cape we continued our course northward, still keeping on the lookout for whales, it having been several weeks since we had seen any, and I began to feel myself out of danger, not forgetting to wish that we never might see another; but I found that my wishes could not keep the fish from our course, for soon the astounding cry of "there she blows" came ringing from the mast head.

"Where away?" cried the captain, as he sprang quickly to the shrouds. "Four points to the lee bow, sir." "How far off?" "About three miles, sir." "What does it look like?" "It has gone down now, sir, headed to the windward." The captain with his telescope ascended the mast, waiting until he arose to the surface. It was not long before he made his appearance, when the captain pronounced him to be just the one he wanted. The boats then were lowered away, and it was not many minutes before the boat in which I was in was fast to the whale, and in as many minutes more we were thrown into the deep. The boat's line was new, and on getting it wet it became full of kinks. One of these kinks caught in the boat's chock, throwing its contents out with a sudden jerk, and taking a bee line astern of the whale, was soon out of our sight, and we never saw it again. While in this position I thought of my friend, Mr. Shark, and there was no mistake but his chance would have been good for a meal, but we were soon picked up by one of the other boats and taken on board. The captain said we were a pack of fools, and were more fit to handle a musket than attack a whale; but I took no notice of what he said, and thought I had rather be a soldier's horse than a whaleman. The next day we saw more whales, but as the cry came from the mast head I thought my time had come. First I thought I would break my arm, so that I could be expelled from duty, forgetting that we had lost our boat the day before, and the other was not yet ready, therefore when the boats were lowered away we were left on board, so that I had a fair view of the scene. They succeeded in striking two very large whales. After they struck, the whales ran and the boats appeared to fly. I never saw anything go so fast in my life; but they soon began to slacken their speed, the boats hauling up to them, the officers using their lances freely. The monsters soon began to throw blood from their nostrils, striking out in a circle as they generally do when they are captured, turning their heads to the east, making one tremendous effort, and their lives were gone. The signal was now set for the ship. We therefore ran down and took the whales alongside, and after dinner preparations were made for cutting in. They were about seventy feet in length, and produced a hundred and thirty barrels of oil. This was the last of my fishing on board the old George.

We stopped at Juan Fernandez. The inhabitants there number thirteen persons, they being Spaniards who were banished for some offence, from their native land to this island. We got from them some yams, sweet potatoes, onions and dried meats, and in return gave them molasses and bread. We then struck a due course for the Sandwich Islands.

Perhaps it would be well to inform my readers how the watches at night are spent, in pleasant weather, and in order to do this I shall have to pitch upon some watch which I can remember. Upon this occasion the watch was gathered around the main hatch, when Bill Brown was called upon to spin a yarn. "Yes, yes," cried half a dozen voices, "give us a yarn, Bill." He was a short thick-set person, with red bushy hair, and freckled face, with a pair of small grey eyes, peering out underneath a set of massive brows, his nose large and sharp, and his mouth wide, teeth uneven, and a large dimple in his chin. This was Bill's first voyage at sea, and he made all the excuses imaginable, but it was of no use, the story had to come; therefore he at once commenced in the following manner:

Two of us down easters, who knew something about catching yellow pike in Lake Erie, determined to go in for a little higher sport than catching blue fish; in short, after taking into consideration all the hazard there was, we settled down upon the purpose of devoting one day to shark fishing. We called in our council, Captain John Ingram, the owner and commander of a trim built smack, of about ten tons burden; a better man could not have been found in the port of New London, thoroughly competent, from long nautical experience and familiarity with all the reefs along the coast, courageous, systematic and cool. We found Captain Ingram full of genuine fisherman enthusiasm, ready to help us in carrying out our resolutions. He knew just the place to go to, out to Race Point, about nine miles off, at the head of Fisher's Island, where the pesky creatures took many a hook and line, bob and sinker, when I've been down there blue fishing. The captain hove into sight early on Wednesday morning; we hailed him from the verandah of the Pequot House and brought him along side the dock. To pick up the fishing traps and shoulder the basket of grub, was the work of no time at all. We had invited a gentleman from Northampton to join us. The invitation being readily accepted, our party was then complete, three amateur fisherman. Captain Ingram and his first mate, old fishers, or rather whalers. The captain said there hadn't been no fishing for sharks, but he guessed he was rigged to try it on with 'em. We thought so, too, for we found him provided with a whale lance, a huge gaff hook, and a line just the size of a bed cord, well tarred, with a hook a trifle smaller than an ordinary pot hook, made fast to the line with several strands of brass wire, closely wound. We left the dock with a stiff breeze, but outside the river Thames it fell off to a dead calm; but the captain allowed that, wind or no wind, he was bound to have a shark; so he and the first mate manned the oars and pulled away under the boiling sun, while we land lubbers crawled under the shade of the sail and snoozed. The tide, setting strongly seaward, was in our favor; and about two hours' pulling brought us to the head of the island, around which the tide runs at ebb at the rate of six knots an hour, and where the ocean surf rolls and dashes on the reefs with its ceaseless roar. The shout of the captain, "Over with your anchor there," roused us at once, and starting up, we found we were on the fishing grounds. As we had but one shark line, two of the party baited smaller hooks and threw them over for flounders and porgies, although the interest of all of us was concentrated on the bed cord. Joe Smith took his station first, with the understanding that we were each to catch a shark before leaving, and Captain Ingram baited the hook with half a fish, and over it went with about a pound of lead attached; about forty feet of line was paid out when the lead sounded the bottom. Joe Smith found shark fishing a good deal like Jim's account of bass fishing off the pier, plenty of glorious nibbles, with the hook skinned, and now and then a sea weed. "Hold on," says the resolute captain, "and you'll have him yet." A few moments after, a half stifled "by jings" escaped the compressed lips of Joe, and looking around we saw him pulled hand over fist, with his feet braced against the stern of the boat, his eyes sticking out, and his teeth firmly set. "Hand me the lance," cried the old skipper. "Heave away on her, my lad," shouted the captain, now as excited as any one. "Keep away from the line and let me play it out," responded Joe to the offered assistance. A few more pulls, and a dark object was seen rising slowly up near the stern, when in an instant down went the lance from the hands of the captain, whiz went the line thought the hands of the shark catcher, while a stream of blood, rising to the top of the water, proved that the captain was right in saying that he had given it to the brute. A steady pull brought him up again, exhausted from the loss of blood, but sure enough there he was, a genuine shark, a regular tiger shark; at any rate, he turned over on his back and opened his huge jaws, set round with a double row of long white teeth. We felt the blood crawling through our veins, and thought at once of the picture in the Bible, of Jonah diving into the whale's mouth, and of the pictures of the heathen mothers throwing their infants into the crocodiles' jaws, and we thought how easy it would be for that chap to take a year old baby at a gulp. While this was running through my mind amid the greatest of shouting and hustling from all hands in the boat, and thrashing and diving of Mr. Shark, the captain was lunging the spear into his sides with a hearty good will, accompanying each thrust with a grunt, and saying, "Oh, you dog, I'll pay you, you ugly beast," while the shark would roll up his wicked eye at the captain, looking as if he thought within himself, "yes, old salt, and if I had you over here, at fair play, I'd make the meat of your leg in a hurry." As it was, the shark was no match for the captain. After fifteen minutes hard struggle he was laid out and towed forward, and made fast to the bow. Old Northampton tried it next. "I guess," says the captain, "the rest is driv away just now, but there is enough of 'em hereabouts, I know." We had about an hour's good fishing for flounders, when all at once not a bite was to be had. "There, boys, the dogs is around again," says the captain; "that's a sure sign, when there is no bites the sharks are nosin' round." Right again for the captain. Stand clear. Hand in your lines. Where's the lance? Lend a hand here. Captain notified to clear the deck for action. "What a lunker!" ejaculated the excited but firm old commodore, and crash went the lance into his jaws. The next blow was well aimed, and hit him in the *withals*, as the captain allowed. This shark measured just nine feet lacking an inch, and weighed two hundred and fifty pounds. I was fortunate to get a strike with the cast. Here was fishing for you. I had pulled out a fifteen pound pickerel, but never had before played a three hundred pound shark on the end of a bed cord. The skin was raked off my hands when the line was whistling through, and proved what the captain called the blasted grit of the pesky critters. This shark died hard, twice breaking away from the lance and the gaff hook, and made the great deep boil like a kettle, but it was no use, he only crimsoned the water with his own blood. At every struggle he measured an inch more than the last. "Well, boys," says the captain, "we've done enough for one day. I've had my revenge on the pesky sharks, let's hoist 'em in and make sail." It took the united strength of five men to drag them into the boat, though we took them one at a time. We had a fine breeze homeward, and landed at the dock at six o'clock. After Bill's story was complete there was a loud hurrah, and eight bells run, the watch was called, and we returned to our berths.

At length the Sandwich Islands appeared. The pilot came on board as we were winding our way through the narrow passage into the bay of Honolulu. The anchors were cast, the sails furled, and we were once more sheltered from the boisterous ocean. Our ship was anchored opposite the fort, her head pointing towards the mountains which overlooked the bay, on the sides of which appeared herds of wild cattle; in the distance they appeared no larger than goats or sheep. Many of them are caught with the lasso, to supply ships with beef. On the right of us lay the United States frigate *America*, and on the left lay several other whale ships. On the next day after our arrival my watch was permitted to go on shore while the remainder of the crew were making preparations for taking in a supply of water. As my feet struck the beach I was caused to reel to and fro like a drunken man, my feet having not yet lost the motion of the ship. But as I made my way towards the town, which was a short distance from the landing, I became more steady, and could make better headway in pacing the bogs than when I started. On entering one of the hotels every countenance looked strange to me, and I thought I was completely out of the world. Finding nothing here for a pastime, I and two of my shipmates resolved to take a ride out of the town to see a little of the country. Three horses were engaged, and we mounted the saddles without delay. Our guide took the lead and we followed in the rear. We found that a good portion of the land was in a state of cultivation. The prospect before us was beautiful, the land generally rising from the sea shore. We had a distinct view of the country for many miles around.

We rode about twelve miles into the country, and coming to a little settlement we dismounted from our horses and visited some huts which were occupied by the natives. They were very hospitable and having no chairs kindly invited us to one of their mats. They commenced a conversation with our guide, none of which we could understand, but he told us they were enquiring for tobacco. We gave them what we had, a portion of which was soon made use of in the following manner:

They formed a ring, the oldest persons among them in the center, and I saw that they were about to take a social smoke. The pipe was first passed to us, but as we refused, the person in the center commenced, and taking about three drams, swallowing the smoke, passed it on to the next oldest person, and so it was passed around the ring, and they appeared to be much pleased with our company, and invited us to dine with them, but their mode of eating we had not been accustomed to, so we declined the offer. A large calabash of poyce took the place of the old man in the center. This fruit is prepared from plantain and bread fruit, mixed with the milk of cocoanuts; when prepared it resembles a dish of mush. They all eat out of one dish, using their fingers instead of spoons. This, together with raw fish, constituted all they had for dinner. After the meal was over they invited us to partake of some water melons, which we readily consented to. We tarried with them about three hours and then retraced our steps to town. We saw some very aged persons on our way, and I think they were the oldest persons that I ever saw. We enquired if they could tell us how old they were, and as near as we could find out, some of them were upwards of a hundred and ten. This reckoning they kept by the moon, cutting a notch on a stick every moon; this stick was kept until notched twelve times, then carefully laid away and a new one taken, and by so doing they got the number of years they lived.

On reaching the hotel from whence we started we partook of some refreshments and then returned to the ship. As I lay in my berth that night

I thought my chance of escape was slim. There were a great many policemen on the island, and I learned that the natives themselves would traverse the island for days and weeks in order to catch a runaway. But, thought I, a faint heart never won a prize, so I was to put my plans in execution. My next liberty day on shore, I bargained with a man who was English by birth, to secrete me in his dwelling until the ship had gone. I must inform my readers that a watch is kept through the night by one person when a ship is in port, each man taking his regular turn. I therefore made my friend acquainted with the hour when my watch would arrive. This was the time when my clothes should be taken on shore. The bargain was made, signed and sealed, and I knew unless my friend proved me false in order to obtain the paltry few dollars which I knew would be offered for my recovery, I should be safe in making my clearance. I had not lisped a word to any one on board, but appeared to take just as much interest in the ship's affairs as before. It was not long before my night watch arrived, and at the appointed hour came my friend noiselessly paddling his way like a thief in the night, towards the ship. I had procured a canvas bag for my clothes so as to leave my chest unsuspected, and as he came along side I lowered the bag to him. He took it and returned to the shore without being discovered. In order to get my clothes from below without detection, I had to plan a scheme, for I was assured it would be almost impossible to catch so many men sleeping all at one time without some medical aid, so I administered them a small quantity of laudanum in the following manner: a bottle of brandy was procured and the brandy and laudanum were mixed, after which I treated my shipmates and they partook freely of the delicious beverage. Visiting their berths before I removed my clothes I found that the liquor had the desired effect; every man of them was sleeping most profoundly, and I began to think that I was quite a physician, laughing in my sleeve for I dared not laugh aloud. I returned to my station on the deck, and I must admit that my appetite craved a portion of the brandy, but for fear of being caught asleep on my watch I dared not indulge; but my watch drawing to a close, and as the day was dawning, the officers one by one made their appearance on deck. Everything was right on my part, and I appeared to be one of the most innocent of mankind. The mate was first on deck, and making some remark about the weather ordered me to call all hands, and as I thought of the laudanum I dreaded to make an attempt to arouse the sleepers, but after hallooing, and shaking them enough to awaken the dead, I succeeded in bringing them to a state of sensibility, and they slowly made their appearance from below, yawning and rubbing their eyes, and declaring they had never slept so soundly in all their lives, my readers may imagine how I felt. My eyes might have been seen to laugh, while my face was as straight as a deacon's. Not one of them suspected that the brandy was the cause of their sleeping so sound.

After breakfast, part of the men went on shore on liberty, and the next day it was my turn again, as I heard that this was the last, I resolved not to return to the ship, and when all hands were mustered that night to see if any were missing, I think I might safely say there was, for I was snugly stowed away in a dark room in one of the hotels in Honolulu. A reward of thirty dollars was offered for me and the search commenced, but my friend proved true, and they had to put to sea without me. Two days after their departure, as I was strolling along one of the streets, I was accosted by one of the police who asked what ship I belonged to. "No ship," I answered. "Then show my your passport," said he. "I have none," I replied. "Then you must go with me," he said. I thought I knew his meaning, and endeavored to make my escape, and should have succeeded had it not been for a small whistle which he had in his pocket, and on his blowing this I was surrounded by about a dozen native policemen; so rather than have a broken head I yielded the point and went with the officer and was soon under lock and key, safe within the walls of the fort. I enquired of the captain what was my offence that I was imprisoned within the fort. He said I was not charged with any crime, but it was against the laws to run away, and this would be my boarding place until I joined another ship. I found my residence much more agreeable than I had expected. At the expiration of five days I joined the bark *Rhone*, from Australia, commanded by Capt. Dennis, and bound for China. I now found myself under the English flag and once more upon the ocean. The first of my exploits on board occurred one morning soon after our departure from the island. Our provisions were not sufficient to satisfy our craving appetites. This I could not stand, so I told the boys I would take it upon myself to mention it to the captain. I did so and without speaking a word he dealt me a blow with his fist. True to my colors I returned the compliment, but I was soon seized by two of the officers, and the captain pounded me to his satisfaction, then sent me to the mast head until he saw fit to call me down. I remained there from eight in the morning until two in the afternoon. As I sat there at the mast head viewing the pond of which I appeared to be the centre, and thinking of the past, I remembered that I had not signed the ship's articles. This was consoling to my mind, and I was resolved not to go any farther than China with him. Notwithstanding the abuse he gave me, the amount of food was increased and we had a plenty. We called at an island on our passage, the name of which I cannot recollect; the natives came on board and brought with them such fruit as their climate produced. We tarried with them about two hours and then went on our course.

Everything went on smoothly until we reached the Chinese sea, when we were chased by a small sloop which proved to be a pirate. She had chased us about three hours and was now within hailing distance; she ordered us to back our main yard and heave to. This our captain refused to do. There were two guns on board of us and during the chase we had loaded them to the muzzle, together with some muskets and pistols. Every man, though somewhat frightened, stood ready to discharge his firelock at the word. The night was very dark and prevented us from seeing their manoeuvers on board the sloop. We were expecting a broadside from her every moment, but our captain thought the first shot was the best, so we gave them the contents of our arms. She fired about the same time, but I rather guess our six pound pills were what they little expected. She rounded to and gave us a second broadside; she was now astern of us. One gun was then run aft and we gave them a second charge. She fired the third round at us, but now was too far astern to do us much damage. We kept straight on our course leaving them to their own destruction. The captain said he rather guessed they were short of ammunition. We now tried the pumps to see if any of her shots had boarded us, but found no leak; some of our mizzen mast shrouds were cut away, the spanker boom damaged, and the mizzenmast was slightly shivered. Whether it was owing to the darkness of the night or they not having experienced gunners on board, I cannot tell, but this, together with a few bullet holes in our canvas, was all the damage we received, and in about twenty-four hours we were safely anchored in the harbor of Hong Kong. As soon as the sails were furled and the decks cleared, I began to think of the abuse I had received on board, and I at once denied duty. This excited the captain, and he threatened to punish me. I pointed to a man of war and told him that when a ship was lying in port that a foremaster's chance was as good as a captain's, and he had better let that job out. Our conversation reminded him that he was not at sea, and he returned to his cabin without putting his threats into execution.

The next day I went to the American consul's, and told him that I had been abused on the passage and I wanted my discharge. I told him I had never signed the ship's articles, therefore I was entitled to my liberty. He promised me that he would visit the ship at ten o'clock the next day. I then returned on board and awaited his arrival. At the appointed time his little boat might be seen rowing towards the ship with the stars and stripes flying at her stern. After the usual compliments with the commander he made known his business, and as he made use of my name, the captain's face changed color. He declared to the consul that my name was on the articles, which he produced. The men were mustered and the roll called, each man answering to his name, until Thomas Smith's name was called, to which no one answered. He affirmed that was my name, upon which the consul produced my papers, which I had given him the day before, giving a description of my person, together with my name as signed by the Custom House collector in New York. Upon reading my name to which I readily answered, the captain looked thunder struck. The consul then ordered him to blot the forged name from his articles, and said if he was ever caught in the action again, the law should be visited on him. I was cleared from the ship according to the consul's orders, and taken on shore before he left.

Being once more free, with a written discharge, and free to go or stay just as I pleased; after taking all things into consideration, I resolved that if I could find a place to work at my trade I would stay rather than traverse the sea. I put up at the Hong Kong Temperance House, and the next day visited some of the work shops. I found that most of the labor was performed by Chinese, for which they received but a small compensation, their wages not amounting to more than eight dollars per month. In one of the establishments I was offered fifty cents per day, but this was not sufficient to pay my board bill, unless I consented to live with the Chinese; but scorning the idea of rat soup and rice, I declined their offer. Notwithstanding my repugnance to a life at sea my situation compelled me to embark for another voyage. My choice lay between a whaler and a man of war, but owing to our engagement with the pirate, on our last voyage, powder and shot became offensive, so I made choice of the former, and signed the articles for a cruise in the Polar Sea, on board the ship *Champion*, commanded by Captain Waterman. I had promised myself never to be caught on board of a whale ship again, for I had a strong dislike of attacking the sea monster, a peculiar dread of being thrown from the boat into the sea, being unable to swim, and a mortal fear of being swallowed up by sharks. When I thought over all the danger I was to encounter my heart sank within me, but crowding down the inner man and putting on a good outside appearance, I went on board.

The anchors were weighed, the sails spread to the wind, and soon we were winding our way out of the narrow passage, into the broad waters of the ocean.

The darkness of the night soon hid the land from our view, and the wind which was in our favor, swept us swiftly on our course. We had a pleasant passage until we reached the high latitudes, then the weather was colder and we experienced many severe snow storms, together with hail and rain, and for several days the ship was covered with a sheet of ice. I can assure my readers, that the climate was different from that of China. In the southern latitudes the look-out at the mast head was relieved every two hours, but here in this cold region, thirty minutes was all a person could endure. The probability is that most of my readers are acquainted with the enjoyment of placing themselves on a cold day before a roaring kitchen fire, but the men who cruise the Northern Seas do not enjoy this privilege. The storm must be endured without smelling the fire, not because of the expense, but men who hug the fire cannot endure the cold as well as those who never see it. The only method of warming the north cruisers is by exertion, scuffling, wrestling, &c.

A few days further north brought us in contact with the icebergs. At their first appearance, the captain thought best to change our course, but on scanning them with his telescope, he perceived that the sea appeared to be clear beyond, and not wishing to lose any ground, kept pressing our way towards them, and in one hour more we were enclosed between two mountains of ice, and the water which appeared clear to our view was gone, and there was nothing to be seen but ice for miles around. The very

name of icebergs ever sends through me a thrill of dread; I almost feel the chillness, as if I had entered the chambers of the dead; but how much more deeply must these feelings pervade the minds of those who, on the restless bosom of the ocean, approach these frigid floating mountains, even in mild latitudes, when solitary they sail towards a genial sea, spreading around them a mantle of fog in the night, and gleaming in the morning sun like pyramids of amber. They are singly objects of dread and danger, but when these mighty monarchs of the ocean are gathered in clusters and surrounded by thick ribbed drifting ice, and night shrouds the dangerous rovers, and polar coldness and raving tempests unite their strength, then we may call icebergs terrible.

We remained in this position for three days, when, by the change of the wind, the iron-like grasp of the bergs was loosed, and with the exception of getting the paint rubbed off from the ship, we received no damage. We were now in the sea of Kamtschatka, and as it was early in the season, and we were short of wood, we anchored in the St. Lawrence Bay, and took on board a supply of north-west pine. One fine afternoon, after this was accomplished, the captain permitted five or six of us to go on shore. We took a straight course for the pines, and had not traveled far before we struck a bear track. Having our axes with us, we resolved to trace him out, and following this track it was not long before we came to the mouth of his den. The entrance was large enough, but there was no person brave enough to explore the interior, and fearing that bruin would commence an attack upon us we switched off from the track.

About two miles from the beach we came to an opening, and a mile in advance of us could be seen a stream of fresh water. Wishing to know whether it was salt or fresh, we bent our steps forward, and on our arrival our hearts filled with joy when our eyes beheld the large quantities of salmon with which the pond abounded. We knew that some of them belonged to us, but how to get them we knew not, as we had no lines or hooks of any description; but it was plain to be seen, by the action of one of our shipmates, that he had devised a plan which he put into execution. Taking from his back one of his shirts, he invited us to follow suit. The seams were ripped apart, and fastened together with strings made by tearing one of his shirts into strips. Loud roars of laughter burst from every one, at this novel method of making a seine. Sinkers were attached to the bottom, and the cloth cut full of holes, when our seine was about complete. But where were the lines to draw it by, was the next thought; but taking the comforters from our necks and each man sparing one suspender, we had line enough and some to spare. By casting our little seine three times we had quite a supply of fish, and as night was fast approaching we thought it best to return to the ship. As our boat reached the side and the fish were passed on deck, I think I might say with safety that I never saw a person's face light up with joy as did our captain's. After enquiring where and how we got them, he said we might spend the next day in the like manner, and he would accompany us.

Before I proceed any farther, I must inform my readers of the habits of the polar whale. These monsters are often found early in the morning, before their slumbers are disturbed, close to the shore, lying in not more than five or six feet of water, with the spray of the sea washing over them, and appearing in the distance like a huge rock. Therefore it was customary among the whalemen to cruise around the bay with their boats every morning, in pursuit of prey.

After the search was accomplished, without success, our boat was hauled upon the beach, and we again bent our steps towards the pond of salmon. On reaching the spot where our old seine lay which we had prepared the day before, the captain burst into a fit of laughter, and said that surely necessity was the mother of invention. Having a good supply of fish hooks and lines, and being better prepared than we were the day previous, we had great success, and, long before the fall of night, had as many fish as we could possibly carry. As we were slowly moving on towards the boat with our burden, our friend bruin saluted us and it was plain to be seen that he had declared war against us. His first appearance sent a thrill of terror through our veins, but recollecting that there were three rifles in the company, together with a couple of axes, our nerves became steady, and we prepared for action. Taking care to keep bruin in the distance, two of the rifles were discharged, but not striking him in the right spot he still came forward, and when within twenty yards of us, the captain discharged his piece. Bruin gave one tremendous groan, and then fell to the ground. Giving him one or two blows with the butt of an axe, and at the same time keeping clear of his grasp, we left him in his flurry, and he soon breathed his last. As the fish was all we could manage, we bent our steps to the beach, and loading our boat we rowed for the ship. After taking supper we amused ourselves by talking over the events of the day, while the remainder of the crew, which was not with us, got completely out of patience on account of being left on board, while we had spent the day in sport. I can assure my readers, that I had not spent such a happy day since I left Europe. Night was now upon us, the crew had all gone to their berths except myself, and as it was my first watch I commenced pacing the deck. It was still and frosty, and nothing could be heard but the roar of the billows as the water dashed in foam against the huge rocks at the west end of the bay. My thoughts ran swiftly back to friends and home, as is not uncommon for the young mariner when passing his first year at sea. All the transactions of childhood come up in his memory; very often the tears of anxiety and affection are seen stealing down his cheeks, and more especially as he thinks of his parents, who watched his opening faculties and who are probably sorrowing for their absent child. I can assure my readers that such solemn thoughts soften the hardest hearts, and cause the boldest of men to shiver. But suddenly thinking of the immeasurable distance which separated me from home, I thought it useless to borrow trouble, so I shook off the cloud by taking a smoke, and upon looking at the time piece which was erected in the cabin, I found it was time to call the second watch, and after arousing the sleepers, I retired to my berth, where I slept soundly until morning, when I was aroused by the usual cry of "all hands ahoy." After breakfast, the boats were manned, and we started our cruise around the bay, and liking the idea of sailing quietly around, I felt at home, and was much better pleased with the Arctic fishing than that of pulling one's life away in the middle of the ocean, chasing a sperm whale, which would run faster than an express train. A six hours pull is nothing in the scorching sun of the Pacific, and returning often without oil enough to fill a lamp, and with our faces tanned as dark as that of a South Sea islander. But I must not dwell upon this subject, as my readers will be anxious to know how the present forenoon was spent. We saw no whales, and wishing for some recreation, we repaired to the place where our friend bruin was left, and finding that he had been unmolested, and having plenty of help, we attached a rope to his neck, and by the strength of twenty-four men he was drawn to the water, the rope was fastened to the boat, and in this manner we conveyed him to the ship, and before the close of the day we had a good supply of pure bear oil.

On the following morning, in our excursions round the bay, our eyes beheld an object we had been in pursuit of. One of those huge fellows lay quietly sleeping, amid the rocks on the distant shore. Paddling noiselessly towards the prize, a pair of harpoons were buried in his side before he was aware that the enemy was upon him. Now kicking, and rolling, and struggling for life, at getting depth of water to float his carcass, he struck his course for the mouth of the bay. In the meanwhile taking care to keep our boats at a reasonable distance, but holding the line, our boats seemed for a few minutes to fly. But they then came to a stop, the whale faced the boat and slowly advancing towards us, and we saw by his manoeuvers that he was about to show fight. His huge flukes were raised above our heads, now bringing them down with a tremendous crash, splashing the water for yards around him, then suddenly diving, bringing the line perpendicular, the officer of the boat fearing that he had sunk to rise no more, as this is frequently the case. Our line was already run out, which was a hundred and sixty fathoms in length, and the line of the second boat attached, and yet he sank; but after sounding the length of two lines he began to ascend, and as he did so we gathered in our line, which the reader will perceive was quite a job. After being under the surface for the space of one hour he made his presence known by the sound of his breathing organs, which may be compared to a railroad engine, with the exception that that of the former is more loud, and the revolutions not so fast, when sending the water from his nostrils to the height of fifteen or twenty feet. Our opponent seemed to await our arrival, lying almost motionless upon the surface of the water, but now our officers seemed determined to kill him or perish in the attempt. Hauling the boat until the head struck his side, the officer cheering the boys by words of encouragement, at the same time burying his lance in his vitals, it was but a few moments before the blood came rushing from his nostrils. There is no one that can appreciate the joy that fills the whaleman's heart as he begins to strike out the circle, that I have before mentioned in my narrative. We soon saw that he was making his last effort, and backed our boat a considerable distance from him so as to be clear of his flukes, for the dying stuggles of the whale are the most dangerous.   The whale is sometimes caught playing possum, lying motionless upon the bosom of the sea, and making as if he was dead, when he is only waiting to get a chance at the boat, therefore it is necessary to give him a trial: and to ascertain whether he is dead or alive, a lance is thrown at the eye, and after this is done, if he does not stir we can work around him in safety. A hole was cut in his head in which the line was attached, and four boats stringing themselves along in a row, we commenced towing him to the ship. I must admit the task was hard, but thoughts of the many barrels of oil which lay around his carcass, gave us courage, and we soon got him alongside.

As I have not yet given a fair description of cutting in the whale, I will now undertake to do so. A cable chain is first put around the flukes and taken to the fore part of the ship, the tackles are sent aloft and the falls then taken to the windlass. Stages are rigged on the outside of the ship, which is the officers' station, running a breast net for their support while in the act of cutting. For this purpose, long handled spades are used. The head is first separated from the body, then a hole is cut in the body for the first tackle hook. These hooks weigh about two hundred pounds each, taking two men to handle the block to which they are attached. In order to fasten this hook, a man has to jump overboard upon the whale's back held by a rope fastened to his body. It has been the opinion of many that the whole carcass of the whale is tryed, but this is a false idea; the only part which is tryed is the blubber in which the carcass is enrolled, and it may be compared to the fat on a hog's back. This was taken off in strips from four to five feet in width and about twenty feet in length, the thickness from six inches up to two feet, and is termed by the whalemen a blanket piece, and this is lowered into the hold, which is called the blubber room. When the cutting is completed the carcass is cut loose and floats away, commanded by sharks, birds, &c. Then next comes the head, the upper part of which is all that is saved. From it is taken the whalebone, for which I fear the ladies seldom think to thank the poor whaleman. This bone grows in the mouth and is attached to the upper jaw; it grows in slabs which are about eighteen feet in length and tapering as it grows toward the throat. It is about twelve inches in width at the butt and generally tapering to a point, and about three-eighths of an inch thick, with coarse black hair growing from the outside edge. Before it is taken from the jaw it resembles a sieve, and to give my readers a correct idea, I shall inform them of the food which nature has prepared for these uncommon mouths. Wherever the whale is found the water is full of small squids not larger than a common fly. When the whale feeds he opens his mouth and runs with all the force imaginable until his mouth is filled, then closing his jaws the water is thrown through the nostrils, and the squids swallowed. This action is renewed until his appetite is satisfied.

The try works are now cleared away and preparations made for boiling. The kettles in which it is tryed hold eight to ten barrels each. The blubber is first cut into pieces about two feet in length and six inches square, and by turning a crank it is cut into slices about half an inch thick, when it is ready for boiling. After the oil is separated from the blubber it is bailed into a copper vat for cooling, and from this into casks, then stowed in the ship's hold, where it remains until the ship reaches home.

Now, reader, I think I have given you a fair description of the whale and the manner of securing the oil, but if it does not satisfy you, you must wait until I have taken another voyage, and I will try and describe it more to your satisfaction. The whale produced one hundred and forty barrels of oil, and I must admit that if whales could always be caught and taken care of when the ship was snugly anchored under the lee of some land, I should like the business first rate.

We tarried in this bay ten days, then crossed the Bhering's Straits into the Arctic Ocean. Here we spoke the ship *Charleston*, commanded by Captain Blakesly. Our captain was invited on board. Accepting the invitation, the main yard was backed and a boat's crew with the captain went on board. By changing boat's crews so as to leave a party of visitors on board each ship, they cruised along together and the day was spent in visiting. This may seem strange to those who are not acquainted with a seaman's life, but if there is anything that is cheering to the mariner, it is when beholding a few strange faces while traversing the fathomless ocean. We saw no whales for several days. The weather was clear and cold, while the pale rays of the sun shone upon us, it being visible twenty hours out of twenty-four, setting about ten o'clock and rising again at two, so that the night is nearly as light as the day.

The eighth day after we entered the Arctic Ocean we saw whales, and a dreadful disaster followed. There was a whole school of them and the ocean for miles around seemed to be alive. The mate's boat soon succeeded in striking one of them, and in an instant one of the men was missing. The line had entwined itself around one of his legs, the whale at the same time starting with all his force, he was dragged from the boat without a moment's notice and before we had time to think, he was out of sight and gone. Thinking it was useless to cut the line at present, we hung to our prize, and as the whale began to slacken his speed we commenced hauling in the line. We soon came to the body of our departed messmate, the line was cut and the body taken into the boat, and the monster that had committed the deed went his way. The body was taken on board, shaved, dressed in a suit of the best clothes his chest afforded. He then was sewed in his own hammock, a canvas bag filled with bricks was fastened to his feet, and he was laid upon the main-hatch to await the going down of the sun, when the body should be committed to the deep, there to remain till the last trumpet shall sound and the sea give up its dead. Perhaps it would be well to inform my readers of a funeral at sea. The ship's speed is slackened and a wide board is procured on which the body is laid. It is then taken to the gangway, one end resting upon the sail, the other upon the shoulders of two men. Remaining in this position until the burial services are read by the captain, the plank is raised up by the bearers and the body committed to the deep. I can assure you it is a painful duty for the crew to leave one of their messmates behind who a few hours before was enjoying perfect health, but now by a sad catastrophe torn from our society when we least expected it, which should be a warning to the careless mariner of the many dangers that constantly surround him. But bent upon their pleasures, absorbed by their schemes for transient good, they think it will be time enough to arouse themselves when the danger is more apparent. But how many have beloved companions on the ocean, concerning whom the deepest anxiety is felt? For a long time no tidings of the absent have been heard; days, weeks and months drag heavily on, leaving behind them only hope. A heart full of affection is kept in cruel suspense, and not unfrequently the mind for years is kept in a state more agonizing and more wasting to the spirit than would be produced by the knowledge of the certain death of the departed. But how little is thought of the friends, to which the departed is strongly bound by the ties of nature; and since the beginning of the world what vast multitudes have been deposited in the seaman's church-yard though no tolling bell has called together sympathizing friends, though no green sod has opened to receive them, and no quiet grove invited them to rest beneath its shadows; yet they have had their funeral services, the winds have sung their requiem, the waves have furnished a winding sheet, and coral monuments mark their resting places. Generation after generation have sunk in the dark waters; and now wait the summons of the last trumpet peal; multitudes more will follow them, and go down to sleep beside them. But I must not dwell longer upon this subject, but resume my narrative.

On one of the pleasant days, as we neared the land, a company of Esquimaux paid us a visit, bringing with them moccasins and bearskin coats; one of them was arrayed in a coat made entirely of duck skins, with feathers of different colors. These articles could be purchased from them with bottles of brandy and pieces of hoop iron, but as we were trying a whale at the time of their arrival, their first business was to satisfy themselves with a supply of scraps, which are considered by them a delicious dish. They thanked us very kindly for the permission of dining out of the scrap tub. As I stood close by I had the opportunity of observing their manners. They appeared to be highly delighted with the treatment they received, and I thought made an excellent meal. Keeping my eye upon the one that wore the duck skin coat, who had been walking around the deck staring at every thing, as if wonder struck, and not forgetting to pocket all the small pieces of hoop iron which had been thrown aside by the cooper, and, as he thought, taking care that no one saw him; at length coming to the large iron ring-bolt which was through and clinched on the other side of the deck, he came to a full stop, and looking round to see that no one observed him, he seized hold of the ring, and I presumed had made up his mind to steal it, but after giving it one pull, with all his might, he thought it too solid to try again, and with a disappointed countenance gave up the job. His next wonder was the ship's bell, looking under, over, and around it, and I saw by his actions that he was trying to form an idea of what use such a thing could be for. Just at this time the cabin boy came forward and struck four bells, and as he did so, the object of my amusement gave three or four loud screams and ran to the spot where the rest of his fellow beings were standing, which set the whole of the crew into a fit of laughter. Had I not afterwards learned that the fellow was a fool, I should have given the Esquimaux a poor reputation, but as it is I shall give them their just dues. They are people small of stature, foreheads low, with straight black hair, and swarthy complexions. Their time is spent in hunting bears, seals,

&c., the skins of which are made into clothing, using the sinews of the whale for thread. They are very often found catching whales, and when so employed a large number of canoes go in company. Their lances are made of bone, and if successful in killing a whale, a portion of the meat is used for food, but as I am not much acquainted with their customs I can give you but little of their history.

We spent the summer season in the Arctic Ocean, and returned to Mohee with six hundred and fifty barrels of oil. The time of our departure from China until our arrival here was seven months, and during that time we had suffered many hardships, privations and dangers, and I think I may say with safety, that it was the severest seven months that I had ever experienced. Notwithstanding this, I had become more reconciled to the seaman's life, and my desire for seeing more of the world had increased, but as I had only joined the captain for one season, the time had now expired, and I was entitled to my discharge. I then joined a sperm whaler, which was going to cruise around the Pacific, and as she was about ready to sail I spent but little time in Mohee. The third day after my arrival, the ship *Empire*, in which I had enlisted, might have been seen looming her way towards the fathomless ocean, and I now found myself mixed with a class of men to whom I was an entire stranger. It was hard parting with my old shipmates, to many of whom I had become much attached, but as it does not take long to get acquainted on board of a ship, I soon felt myself at home, and as every one had some new story to tell me, the night watches for a few weeks passed off finely. In return I had to tell them all the stories I could think of besides some I had borrowed.

I am sorry to inform my readers, I was doomed to encounter the greatest trial that had ever befallen me; and to give a true sketch of the scene, I shall have to mention a few words concerning the mate, whose name was Nye. He had been on board the Empire twelve months, during which time other officers had caught several whales, but he had been unsuccessful, and was looked upon as one who did not understand his business, and unworthy of the positon which he held; but he had declared that if ever his boat was fastened to a whale he would kill him or die in the attempt. The day on which the incident occurred to which I refer, was one of the most pleasant days I ever witnessed. The weather was calm, and not a ripple could be seen upon the bosom of the ocean, except a small portion which was disturbed by the rush of whales. The boats were lowered away, and two of them succeeded in getting fast, and the mate's boat was one of them and the third mate's the other. The latter soon captured his prize, and in the course of two hours was safely alongside, but the one to which the mate had fastened had run so far that nothing could be seen of them. A slight breeze had sprung up, and with it came a fog. The ship was now pointed in the direction where the mate was last seen, but was not able to make much headway with such a burden at her side. The fog became so thick we could scarcely see the length of the ship. The four pound gun was mounted and fired unceasingly, and, as the night was upon us, three globe lanterns were set at the mast-head; the men shouted and the ship's bell was rung, but all to no purpose. Our mate and five of the crew were gone, and no one could tell whether they were dead or alive. At length the captain gave orders to cut loose from the whale, which we were towing. Here was eighty barrels of sperm oil, which we had run the risk of our lives to obtain, set afloat. It was no pleasant duty to let it go, but when the orders came the work must be done. More sail was crowded on, and as we had a good breeze we speedily sailed in pursuit of the lost, firing the guns often. Not an eye closed in sleep during the night. I never shall forget with what anxiety the crew watched the dawning of the day. Tears of anguish stole down the cheeks of the captain as he rapidly paced the quarter-deck. At length the morning broke, the fog disappeared, and as the sun made its appearance, and the horizon around us was once more clear, every eye was strained, but saddened were all faces, as the captain scanned the horizon with his telescope, and with a throbbing heart declared that nothing could be seen. Our hope of ever seeing them was past, but the search was still continued, and on the following morning we saw a ship to the windward of us. Our colors were immediately set at half-mast, as I before said, as the signal of distress. She was pointed towards us, and in a short time ran across our stern, and speaking to us through his trumpet, the captain enquired what the trouble was. After telling him what had occurred, he said he would come on board. He had seen nothing of our lost boat, but deeply sympathized with our captain in his loss. He stayed with us about an hour, and then took his leave, promising, as he did so, to keep an eye out for the unfortunate boat's crew. We then braced our yards to the wind and the search continued. Two days later we found the monster which had been the cause of our calamity. the mate's name, together with the name of our ship, was inscribed on one of the boat's paddles, from which we were satisfied they were still alive, but thinking it useless to continue the search longer, we took care of the whale, which yielded us fifty-five barrels of oil.

We cruised in search of whales for three months, then started for port, having taken about two hundred and fifty barrels of oil. It was the intention of the captain to visit some of the Marquesas Islands, where we at length came to anchor, in the bay of Neukeheva. In the first place I must acquaint my readers of the joyful news which the pilot brought, that our lost boat's crew had called at the island, where a part of them yet remained.

To inform you correctly of their adventure, I must draw your attention to the time of their departure from the whale. They had lain by it twenty-four hours, then thinking that we should never find them, they pointed their boat toward the islands, which were about fifteen hundred miles distant. They had only four gallons of water and twelve cakes of sea biscuit, and were obliged to stint themselves to a small quantity each per day, scarcely sufficient to sustain life. Several days had passed away in this manner, their strength had dwindled away, and they could scarcely guide their boat on its course. Some of them would have soon perished had it not been for the good fortune of catching a porpoise. This gave them strength, and after the lapse of thirteen days they landed on Robert's Island, where they were kindly treated, and taken to the huts of the natives. After tarrying with them about six days, they filled their boat with fruit and started for Neukeheva; where they arrived in safety. The mate and two others had joined a schooner and left for California, and we afterwards learned that they were capsized and lay on the keel three or four days before they were rescued; the other three came on board and resumed their places as before. This island was governed by the French; a man-of-war frigate was lying at anchor, and there was also a regiment of soldiers upon the island. The natives resembled those of the Sandwich Islands. We got from them some fruits, and after taking on board a supply of water, we started for the Society Islands, and on the passage called at an island on which the inhabitants were very savage. They seemed to be at war with the other islanders, not allowing any one to land in their dominions; but as we had on board twenty-four live hogs, which lived entirely on cocoanuts, and the captain thinking he could trade cheaper on this island, resolved to venture, and for this purpose volunteers were called, when I consented to be one of the number. Each man was armed with a brace of Colt's revolvers, and we took our leave from the ship, and pulled for the bay. On our entrance we soon saw they were a wild race of beings, and as savage as had been represented. On our nearer approach a warlike whoop was given by the chief, and in a few moments more the beach was completely lined with a barbarous set of people, and I must confess that when their arrows were pointed at us we were somewhat frightened; but on presenting to their view some pipes and tobacco, it had the desired effect—their arrows dropped to their sides and peace was declared. The chief then swam to our boat, and after learning our errand, returned and in a short time our boat was well supplied with cocoanuts. They were brought to us in bunches, which they held in one hand while they swam with the other; and on receiving their reward they licked our hands and pronounced us "mi-ti-ki," which signifies good to eat. The most that attracted their attention was our fire-arms. The chief requested us to discharge one of our barrels; he was very much delighted, and offered to give us his little boy in exchange for one of them. We refused to comply with his request, upon which he returned to the beach with a sorrowful countenance. After our trade was completed, we renewed our course for Otaheite, which we reached in a few days. Here we took in a supply of oranges and limes, of which the island abounds. This island was guarded by a French garrison, and the natives were kept very strict. All persons had to be at their homes at eight o'clock, with the exception of those whose business compelled them to be absent. During our stay here, which was but a few days; I witnessed one of the most shocking spectacles my eyes ever beheld.

One day after our arrival a brig made her appearance, the pilot was soon on board of her, and after bringing her safe to the anchorage, her captain might have been seen making his way to a French line-of-battle ship, and we soon saw by the manoeuvers that he had brought them some news. Her ports were closed, and she was soon under way. We afterwards learned that the brig had been attacked by a pirate, which he had sent the battle ship in search of, and was now anxiously waiting her return. Two days later the French cruiser returned with their prize; the rogues were caught, and were now waiting the verdict of the court martial. After

Hughes

having the sentence pronounced against them, that at the hour of sunset they must prepare to meet their awful doom, and as the sun was setting behind the western horizon the inmates of the harbor seemed still as death at the sight of the bold rovers, as they stood in line across the man-of-war's deck. Their hands were fastened behind them, and, as the ropes which led to the yard-arms were fastened around their necks, they stood erect, with the undaunted boldness and careless indifference at the awful fate that awaited them. A short address was delivered to them by the commodore, which they heeded not. The orders were now given, and twenty of these wretched victims could have been seen swinging at the yard-arm, when the crowd that had assembled to witness the scene dispersed. On the following day we took our leave of the island and proceeded on our way to Pitcairn's Island, setttled by mutineers of the ship *Bounty*. A great many of the whale cruisers call at this island for yams and sweet potatoes, the quality of which is much better than that of the other islands. The inhabitants of this island make use of the English language. They are very strict, and subject to no law but their own. As our anchors were not cast here our stay was but a few hours, and the ship was kept in motion during our short stop.

We next called at an uninhabited island, called Terrapin Island. Ships call here for the sole purpose of getting terrapin, a species of land turtle, where they are found in abundance. While in search thereof, we found a terrapin which was an old settler. He had inscribed on his back dates and names of ships, some thirty years before our visit to the island. Every one had respect for this noble fellow, and we left him unmolested, to roam over his dominions. The first inscription was his title, which was Mathusalah. We took from this island ninety-six, which made us many an excellent meal. I have known these terrapins to live six months on water alone. Unaccountable as this may be to my readers, it is an absolute fact.

We were now bound for the coast of Peru. Six days after our departure from the island, we experienced a heavy gale, and while in the act of reefing the fore-top sail, the sail filled with wind and I was knocked from the yard, and should have lost my life had it not been for the timely aid of a Portuguese, who, in my descent, caught me by the hair of the head with one hand, while with the other he clung to the treacherous rope until I regained my footing. During this gale a large sperm whale made its appearance a short distance from the ship, lying apparently motionless, and by its actions appeared to be laying to during the storm. The captain's eyes shone with unwonted brilliancy, and, I must admit, he looked very tempting notwithstanding the gale. At length five dollars was offered to each volunteer that would attempt an attack, at which the crew shrugged up their shoulders and looked wishful at this monster of the mighty deep; at length one and then another, until five had manifested a willingness to go. One more was lacking to complete the boat's crew, and no one seemed willing to fill the place, but thinking I should be called a coward and a lubber, I thought it best to venture. A pair of life preservers were put around each man, and our boat was lowered to the surface; the ship gave a heavy roll and, before we could get clear, our boat was filled with water, but taking care to keep right side up, the water was soon bailed out, and in ten minutes more we were fast to our prey. With a sudden start he darted forward with unwonted velocity, and our little skiff dashed on with lightning-like speed, first on the mountain then in the valley, expecting every moment to be swallowed up by the huge waves, but after towing us two miles his speed was slackened, when we approached him and soon succeeded in giving him his death wound. After a couple of hours hard rowing amidst the waves of the boisterous ocean, we regained the ship with our victim, and on reaching the deck we were met with a smile and a bumper of brandy, and I think I may say with safety that we were more than one degree higher in the estimation of the captain than before.

After taking care of this oil we again steered our course for Peru, and after a few weeks' steady breeze came to an anchor in the port of Payta. I was now among a class of people entirely different from any that I had before visited. They are a swarthy looking people, filled with superstition and treachery, carrying long dirk-knives in their boots. My stay was only two days here, during which time I visited the Spanish grave-yard. The soil was mostly sand, and on account of the bodies being buried so near the surface, the frequent rains and winds disinter the dead and their bones are exposed to view.

Here I joined a small bark which was bound for the Sandwich Islands. A few days from port, during a calm, while a part of the crew were in the act of bathing, a shark made his appearance, and the alarm was given; but it was a moment too late to keep the ferocious monster from his prey. They all reached the ship save one, he being at a greater distance and somewhat frightened, it retarded his progress, and before he could reach the line that was thrown him, one of his limbs was severed from his body. After hoisting him to the deck the wound was dressed to the best of our ability, but all to no purpose, for in the space of one hour death terminated his sufferings. He was sewed up in his hammock and consigned to a watery grave in the manner heretofore described. A new officer was now to be chosen, and I was selected by the captain to fill the place of the unfortunate mate. Nothing occurred during our passage, with the exception of losing the fore-yard which was done during a heavy storm. On our arrival at Honolulu I visited my old friend who secreted me on my first voyage. He informed me that the ship from which I had escaped had lately visited the port and had sailed about one week before our arrival. Had we arrived one week earlier, instead of my present situation, I should have been placed before the mast. Our bark had sprung a leak, and the cargo was now to be discharged and the vessel put under repairs. Our stay at this port was one month, which was the longest time I had rested since my departure from New York.

Two natives of this island manifested a desire to try a life on the ocean. They first applied to me for a berth. I told them to apply to their king for permission to embark. He refused their request unless we would pledge ourselves to return them again to the island. This we could not do as it was uncertain where our next voyage would be. They next proposed to run away, and after mentioning it to the captain, he gave me leave to do as I pleased so I informed the natives what course to pursue. On the last night of our stay on the island the two fugitives swam off to our bark, where they hid in safety until we were some distance at sea, when they were called upon deck, where suitable clothes were given to them and they already began to look like young heroes. I took them both in my watch, and my first business was to learn them the English names of the ship's rigging, masts, spars, &c., &c., which they learned very readily, and long before the close of the voyage had become very handy.

We were now returning to the Peruvian coast. Up to this time the crew had behaved themselves well and things went on smoothly, save a grudge which the second mate held against me since the mate's death, on account of my filling his place, which office I still held. It is the general custom on ship-board, when an officer dies, for the next in rank to fill his place, but the second mate on board this bark was an apprentice, and his time had not expired yet, and he was not an able seaman, therefore the captain had to choose one whom he thought competent to take care of the ship. I shall have to inform my readers of a law that if a ship is lost while under the control of an apprentice, the insurance money is forfeited. I had learned by the two natives that the second mate was bound to have his revenge, and that he had already made arrangements with the crew for raising a mutiny. I informed the captain of this news, and told him we were the victims of a plot, which struck him with surprise, and for a few moments stood considering what course to pursue. He was not long in thinking of the only chance that would save our lives, which was to take the second mate by surprise, when looking over the side, and throw him headlong into the sea, which was accomplished after preparing ourselves for the attack, at the same time giving the alarm of a man overboard. The boat was then lowered to the surface, and six of the mutineers were sent in pursuit of him. After getting some distance from the ship, it was our scheme to leave them behind. The man at the wheel was ordered to put his helm up, which he refused to do, when a pistol was pointed at him, with the command to "put up or die." He soon took his choice, and the wheel went up. "Square the yards," cried the captain, to the remainder of the crew, who were standing forward, and as yet knew nothing of the scheme. They ran to the braces, the helm was steadied, and every sail was filled, and we speedily left a portion of the mutineers astern. The remainder were then commanded to muster upon the quarter-deck, and to our great surprise they came, falling upon their knees and begging the captain's pardon, and declaring that they had been led away by the second mate, together with some of the men who had gone in the boat. The captain then enquired of those who were imploring for mercy, if they would promise to obey orders, and do their duty as heretofore, to which they readily answered "yes." After giving them a severe talking to, they were divided and the watch sent below. Neither the captain nor myself slept any that night. We never heard of the boat's crew, but as we were near land there was no doubt but they reached it in safety. I can assure the reader the captain and myself had many a hearty laugh over this night's transaction, not forgetting to befriend the natives who betrayed the rogues. We reached at length the bay of Tombas, which was our destined port. We found that the ship fever was raging at this port, and within three days after our arrival the captain was seized with this terrible disease. He was immediately taken on shore and placed under the care of a Spanish physician. Not liking his Spanish attendants, he sent a line to me with a desire that I should attend him during his illness. I therefore left the bark in charge of the harbor master, and went to his assistance. After the lapse of one month, seeing there was no chance of his recovering his health for a considerable length of time, new officers were engaged and the bark sent home. I watched over the captain until he was able to take passage for the States. I and my two natives, as they were yet with me, and who declared where I went they would go also, joined the whaler which was bound for a cruise in the Okotsk Sea. I now found myself again before the mast and sailing rapidly towards the whaling ground. My two natives were much pleased with the idea of seeing the sea monsters. They had become very handy on ship-board, and received wages. They had learned all the names of the ship's rigging, and could also repeat the mariner's compass, and stood their trick at the helm, when their turns came around, and I must admit that I was very much pleased to hear them tell over their experiences, more especially as they spoke of steering the ship, which they thought was wonderful, to think a man could guide such a noble piece of workmanship, to any part of the world, without a landmark. I often tried to inform them of the art of navigation, and the use of quadrants, sextants, &c., for taking latitude and longitude, but I could never make them comprehend my meaning. They learned the English language very fast, and in a short time became noble seamen. They had always been familiar with the water, paddling around their island in their canoes, and they could swim like fish, and as yet knew nothing of the dangers that surround the mariner. They were strangers to fear, and were always ready to brave any danger, at the word given. They were noble, generous and peaceful, and desired the good will of all men. I may be prejudiced in favor of these two natives, and I have a good reason to be, on account of their having saved my life from the mutineers on my last voyage. I am sorry to inform my readers, that during a northwest gale, the youngest of the fugitives came to his end, in the following manner:

The weather was very cold, which affected them very much, as they were always used to a hot climate, and while in the act of reefing the foresail, his hands had become very cold and benumbed, so much so that he could hardly hold on to the yard. At this moment, the man at the wheel not paying strict attention to his business, suffered the ship to round to from her course. This brought the wind in another quarter, when the sail immediately blew over our quarter, and this unfortunate being was dashed from the yard, and in the descent his head came in contact with the rail, which threw him amidst the boisterous billows, where he sank to rise no more. It would be impossible to describe the feelings of his companion. For a moment he appeared to be deprived of his reason, and would have jumped after his mate, had I not spoken to him in his own language, and forbade his attempt, and it was a long time before he became reconciled to his loss.

On one of those calm days while we were cruising around the Okotsk, a curious spectacle presented itself to our view. A company of seals made their appearance close to the ship. Their heads are similar to that of a dog, while their bodies resemble that of a fish. They would raise themselves up in the water so that their bodies would be perpendicular, and with their heads alone above the surface. We shot at them several times but without success, for they were capable of dodging a bullet as often as we saw fit to fire at them, but as soon as the gun was discharged their heads would pass out of the water, staring at us in the face as if to say, "you can't come it." After trying our skill about an hour we retreated and called them shot proof. Here we cruised about three months, and during this time we captured four whales which produced about four hundred and fifty barrels of oil. We then bent our course towards the sea of Japan, and after cruising a short time we captured a couple of blackfish, and, as it was near night, we lay by them until morning. During the night a boat's crew of eight men lowered a boat and ran away from the ship. They had taken all their clothes, a mariner's compass and a supply of provisions. They had caught the officer of the deck asleep and bound him fast to his seat, and then endeavored to make their escape, but, owing to the noise of the boat's fall, the sleeper was aroused, and finding himself fast he soon applied his knife and gave the alarm. All hands were mustered and three boats were soon in chase of the runaways, who were yet in sight but about a mile distant. We gave them a long chase, but should not have succeeded in bringing them back to the ship had they not turned their course and came towards us, which they did for the purpose of having a combat, knowing unless they drove us back we should be in sight the following morning. They ordered us to turn back, saying that if we did not they would blow our brains out and feed our bodies to the sharks. The captain then made them a fair offer, promising them if they would quietly return to the ship nothing further should be done about it, which they refused to do. We then pulled close to them and undertook to board, but they managed to keep us away by using their paddles, which they did to perfection; but not considering that there were in a small boat, they all gathered to one side and their boat capsized. They now began to cry for quarters, promising that they would return to the ship in peace. We therefore picked them up, tying their hands behind them. Some of the best swimmers jumped into the water and turned the boat right side up and picked up most of the runaways' cargo, when we took them in tow and returned to the ship. These men were then put in irons, where they were to remain until we reached port.

On the following morning preparations were made for taking care of the fish we had taken the day previous. After this was done the captain thought it best to make port; we therefore steered for the harbor of Hong Kong, which we came in sight of in a few weeks, and the pilot came on board of us when about ten miles from shore. As the sun was almost setting, and as the passage to the harbor was very narrow, with huge rocks on either side of it, we thought it best to lay off and on until the following morning. About eight o'clock that evening a dark cloud was seen rising in the distant horizon, and the muttering of an approaching storm was heard. Soon the roar of the tempest broke the stillness of the hour; dim shadows filled the sky, and we knew by the flashes of lightning and the terrible peals of thunder that an awful storm was about to burst upon us. Our pilot, who was a Chinaman, said it was the beginning of a typhoon. Our canvas was taken in, and before twelve o'clock we were under bare poles, the helm was lashed and the ship given up to the winds and the waves, and long before daylight the decks were swept, carrying away part of our bulwarks, cook's galley and water casks, together with the foremast and main-top-gallant mast, jib-boom, &c. The pilot's boat, which was towing astern, was torn to pieces and nothing but the ringbolt left which the line was fastened to, the loss of which caused him to mourn without ceasing. The men who were in irons were now liberated that they might have an equal chance with us of saving their lives. We were all lashed to the rail, expecting that every moment would be our last. We remained in this position three days and four nights without food, water or sleep. It was the hardest storm I ever experienced, and I resolved if I lived to reach port to bend my steps homeward, and never venture again upon the treacherous billows. The storm abated and the clouds began to break away, and after taking observations we found we had been driven more than five hundred miles from our desired port. Our first meal after the storm was bread and raw pork. A good cup of tea and coffee would have relished very well, but as our cooking utensils were lost it was no use thinking of tea or coffee. A jury mast was now put in the place of that which was blown away, and we were slowly moving towards the harbor before mentioned. Had the sea now been alive with whales and our inclinations ever so great for capturing them, it would have done us no service, for our whale boats, seven in number, were lost during the storm. We reached our anchorage in safety and without the loss of life. On the following day after our arrival at Hong Kong, I and my native received our discharge.

It may appear curious to my readers when I inform them that about one-half of the population of this port live in small boats along the sea shore, living mostly on fish and rice. Each of these boats has a small cabin, in which is placed a table, covered with the choicest of fruits which their country produces, placed there for Josh, in English, the devil. When night approaches the candles are lit, and the tables are set in extra style, for the purpose of pleasing Josh. They believe that God is good, and will never punish them, and that all they have to do, in order to be happy after death, is to keep on good terms with the devil. Those who live on land have a room in their dwellings for the same purpose, and fixed in the same style. In their morning worship they may be seen upon their knees, tossing up two blocks of wood until they strike the ground with a particular side up. When this is accomplished, they touch off a bunch of fire crackers, which ends their exercises. During my stay at Hong Kong, I visited the prison, a large building constructed of stone. It was very high, and built in a circular form, containing a large number of prisoners from almost every nation. The white prisoners who were sentenced to hard labor found employment in breaking stone, and the natives could be seen working in companies, upon the road, with fetters upon their feet. When their day's task was finished they were hand-cuffed and fastened in pairs, and driven with a whip back to their cells. For some offences the sentence is whipping. For this purpose they are tied hands and feet fast to a frame, standing in the jail-yard. On the day of my visit to the jail, one of these miserable creatures was fastened to the frame and received twenty-five lashes with the cat-o'-nine tails, every blow drawing blood. Some might think this a hard mode of punishment, but unless some severe punishment was inflicted on the offenders there would be no living among such a set of thieves as Chinamen are. But the most perplexing punishment they have in their laws, is in cutting off their hair, which has been growing from their crown all their lives, and probably would measure five to six feet in length. It is braided in one strand, and is left hanging down the back; and it is not uncommon to see their hair, when walking, drag on the ground. When one is seen without this ornament, we may be assured that he has committed some heinous crime, for which they are despised among their own race.

Some of those who have been punished in this manner, to hide their disgrace, are found wearing false tails, but they are very cautious to keep them clear of the sailor boys, as they frequently give them a yank, when passing in the street, and, if the hair is false, it is separated from the head very easily, and brings to light their dishonor. I should not relate this circumstance was I not knowing the truth, for as I was walking the street one day, one of these ornaments felt a jerk, and as the tail did not belong to the head that had it on, it was left in the hands of the person that gave it the yank. The reader may well imagine what feelings would be produced by pulling off the cloak of iniquity. I have seen some in this country, who, I believe would blush should they by any misfortune be deprived of their cloak. I have seen some that were very thin, almost transparent, and others so narrow that it was almost impossible to keep them together, while some have those whose ample folds cover them so nice that you would never know what was underneath them was it not that now and then a gust of wind opens them a little, and you catch a glimpse of the large bundles of deceit and falsehood which are hugged so tight together that there is room for sundry other articles of the same nature; but if you are not quick of perception, you will never comprehend what diamonds of value shine in the darkness under his cloak, but which the light of truth would sully forever. Therefore you must see that the owner of such a cloak should be cautious, especially in windy weather. But I must not dwell on this subject.

Before I arrived at this place, I made up my mind to sail for the States, but on looking around the harbor I found no vessel that was homeward bound, so I had to take up with the next best chance that offered itself.

A few days later a small brig made its appearance. She had lost two men on her passage, and had stopped here for the purpose of getting others to fill their places. I and my native made just the number, and as there was no other opportunity, we readily accepted the offer, and joined the brig. She was bound for Tombas, a port which I had previously visited. We were followed by a very pleasant breeze, and the little *Water Witch*, for that was the brig's name, plowed the water with perfect ease, and glided over the waves as if sensible of her duty, and the watches passed away pleasantly. The officers were sociable and kind, and it was more like a pleasure voyage than any that I had before been. As we neared the equator we encountered severe rains and terrific peals of thunder, accompanied by lightning. We saw a strange sail in the distance, and as we neared her, for a few moments, she appeared to be on fire. We therefore ran to her assistance, but before reaching her the fire was quenched, but the captain was anxious to know what had happened, and where so much blaze came from, which had been extinguished so suddenly. When in speaking distance we learned that lightning had struck her foremast, and that they had cut it away to save the ship from the flames that had entwined around the mast. Being unable to do them any service, we again bent our course.

When in these latitudes I had the pleasure of beholding a water-spout, which was a most magnificent sight. It was a large body of water descending from the clouds into the sea; but as we were a great distance from it, I cannot give you a correct description. It appeared to be about ten or twelve feet in diameter and a solid body of water. These water-spouts are very dangerous, and if falling near a vessel it is generally drawn under by its force, and swamped in an instant.

The remainder of the voyage was pleasant, though somewhat tedious, on account of the great distance, but we at length came to anchorage at Tombas, and commenced discharging our cargo of sugar and molasses. It was quite a task to remove the cargo in this port, on account of a large sand bar, which prevented the ship from going near the land, and it had to be done with barges. In order to get water for the ship's use, a raft of casks was towed up the river about three miles. The river is long and narrow, emptying itself into the sea, with trees and shrubs on either side protecting it from storms, and it is generally calm and smooth. The village of Tombas is about seven miles in the interior, and there are but a few houses at the mouth of the river, and they are occupied by the harbor master and those who have dealings with ships. It seemed more like a barren island than an inhabited land. This river is noted for crocodiles, and great quantities of them may be found here. While towing our casks up the river, we saw one of these huge animals upon the beach, lying like a log, with his jaws extended and pretending to be asleep, for the purpose of catching flies, frogs, &c. We approached as still as possible, in order to get a correct view of him, but his eye caught a glimpse of us; with one bound he was lost to our view in the bottom of the river. I should judge from his appearance that he was ten or twelve feet long, but those of the smaller kind, which are called lizards, measure from two to three feet, and may be seen at any time lying upon the banks of the river. I must admit that I very much disliked the idea of drinking water from this river, where these animals abounded, but such was the case, and we had to put up with such water as we could get.

The brig was partly owned in Tombas, and kept for the purpose of trading to and from the different islands and places where the owners saw fit to send her. She was now to take a trading voyage, stopping at the Gallipagos and Sandwich Islands. We got at the former islands, two hundred barrels of sweet potatoes, giving them sugar in return. Our stay at this island was but short, therefore I cannot tell much about them or their proceedings, but I will endeavor to give you a faint description of a cloudless evening in the torrid zone. Not a cloud was to be seen in the heavens, and at this time the moon was at its full, glittering like a polished ornament in the deep blue arch above, while countless millions of stars shone with unwonted brilliancy, and appeared like diamonds in the sky, and the galaxy which may be seen in any climate, as a broad zone, passing from north-east to north-west, in these latitudes shines with far greater brilliancy than in any other part of the world, while the zodiacal light, which is always visible in the torrid zone, and which appears a short time after the setting of the sun, something in the form of a cone, illuminating the sky with its brightness, and adding redoubled splendor to the scene. As I stood alone gazing upon the magnificent heavens, I became lost in thought, but as I turned my eye to the fathomless ocean, not a ripple could be seen upon its bosom; its silvery surface reflected every object; it was a mirror of the skies, the pale moon and the stars. From the deep reverie into which I had fallen, I was aroused by the ringing of eight bells, when I retired to my berth, and soon was rocked to sleep by the slow motion of the vessel. I had neve before paid so much attention to the beauties which surround the mariner, and I think it was the most pleasant watch that I ever spent at sea.

The next port we made was Honolulu, and after trading in a similar manner as before mentioned, were ready for the next orders.

This was the third time I had visited this port, but as I had before informed my readers of the habits of the people, my story at present will be short, but you will recollect this is the island from which the two natives escaped, one of which yet remained with me, and the other slept beneath the waves of the ocean. It was the desire of the former to once more tread his native soil and make his friends a visit, but I knew unless he sought pardon from the king before going on shore, that neither him nor the ones that secreted him would escape punishment. For this reason he was hid in the cabin until I should make a visit to the king, and on arriving at the house I asked to see the king, which was readily granted me. In the first place I asked him if he remembered the time when two natives escaped from the island. He answered in the affirmative. I then asked if the natives were returned in safety to the island, if he would grant them a free pardon and give them a passport to go and come when they pleased; after considering the matter for some time, he finally granted my request. I then informed him of the one who had lost his life. This he did not like so very well, but after considering that all were liable to misfortune, blamed no one for the accident. I then left his residence promising to call again on the morrow and bring the fugitive with me. On returning to the brig I told him of the good success I had in my conference with the king, and that on the morrow he was to return with me. I also told him it was no longer necessary for him to remain secreted, and as it was then night we retired to rest, and on the following morning I fulfilled the promise which I had made to the king the day previous. I had informed the native on the way what excuse to make for his leaving the island, such as seeing a part of the world and obtaining suitable clothing for himself. It all passed off finely, and after receiving a slight reprimand from the king he received his papers and withdrew.

The next day was the one set for our departure, and I shall never forget the feelings manifested towards me by this native and his relatives as I took the parting hand. Our brig was now ordered to Wampoa, a Chinese port. Our voyage was prosperous, having good winds and fine weather. On entering this harbor the first spectacle that caught my eye was the largest ship I had ever beheld. She was very long and narrow, and sharp as a wedge, deeply loaded, and appeared like a perfect water cutter. Her burthen was twenty-five hundred tons, and she was called the *Flying Cloud*. Our brig anchored right abreast of this leviathan, and I must confess that I felt ashamed of my positon on board the little craft, as she was nothing but a yawl boat compared with the other. I learned that she was waiting for a crew and then bound for New York. I had been thinking I should like to finish my sailing in such a craft, I joined without hesitation, but in one week from that time I would have given some boot to have been back on the little craft I had made so much fun of. The first night out of port, before we could get clear of the land, a gale sprung up, and we had to cast anchor in fifty fathoms of water. The men were ordered to stay aloft and furl the fore topsail; fifty men were soon on the yard, but as the sail was wet and the wind blew so tremendously, it was utterly impossible to accomplish the order. After trying for two hours we gave it up, and descended the rigging, but were met by the officers, who declared they would shoot the first man that struck the deck. One of the men told them to shoot as soon as they pleased, and it was no sooner said than done, but the bullet passed between his legs and did not damage, and seeing that they would as soon shoot a man as a dog, we thought it best to obey orders. We then returned to the yard where we remained until the next morning, when the wind abated and we were able to furl the sail. The rest of the sails remained unfurled, which took us until noon.

We had now been on board upwards of thirty hours, without eating or drinking, and was permitted to take our first meal, but we scarcely had time to satisfy our craving appetites before the commander made his appearance on the quarter-deck, (he was more fit to be called a hangman,) and gave orders to get the ship under way, which took us until ten o'clock. As the men had not yet been divided, the watches were now chosen and one-half sent below, with the orders that no man should sleep with his clothes off, not even his boots. I must say that the orders were somewhat strict, but notwithstanding this, the four hours' sleep was refreshing. The labor was much harder on board this vessel than any other I had before sailed on. She would spread as much canvas as three common vessels. From the top of her mainmast to the deck, was one hundred and fifty feet, and the main yard was one hundred feet, the centre of it being eighteen inches in diameter, and the other yards and booms in proportion. During the voyage we met with nothing but hard treatment and ill usage. We spoke the *Hornet*, on out passage, which came in sight of us at the dawning of the day, every stitch of canvas that would hasten her speed was now spread and about two o'clock we were side by side. The captains spoke to each other through trumpets, and then handed them to their wives, (for both had their companions along.) We learned that she had sailed from the same port, fifteen days previous to us, but we rapidly swept by them, and were soon out of speaking distance. A few days later we experienced a gale, which lasted forty-eight hours, during which time we averaged nineteen knots an hour, and running under double reefed topsail, with a reefed mainsail and foresail. In pleasant weather it took a man to steer, but now the strength of four men was hardly sufficient to keep her on her course, and as she rolled the water would pour over the rail, and instead of riding the waves, which ran mountains high, she forced her way through them, and the water rushing over her bows at every sea, trembling, surging, joints cracking, masts bending, and yards dipping at every roll of the hulk, and anything but a sea-dog would have hove their ship to, and waited until the gale was over. We lost three men in our passage, one of whom fell from the main-top gallant yard, while in act of reefing. We made New York in ninety-seven days, and as soon as the steam tug was towing us into the harbor, it would be impossible to describe the rapture that filled my heart, to think my wanderings were about to terminate, and that I was soon to rank in the society of enlightened people. I had seen enough of the world to satisfy my curiosity, and I had endured enough of the perils which follow the mariner when cruising over the mighty deep, although I had enjoyed good health, and for sixty months never missed a watch, yet I had become weary of a seafaring life, and having come to this conclusion, there was no place like home.

Yet no abiding home have I,
  No sympathizing friends;
I hope to meet you all by and by
  In that world that has no end.

Surely I have another lot,
  There is some home for me,
Where I shall rest, remembering not
  The dangers of the sea.

I have endeavored to give you a correct history, as far as my memory would admit, of the five years' travel on different oceans. Though this writing may not be as grammatical as many histories we have on record, yet I trust it will not be wholly uninteresting to the reader. Friend, are you sorry that you have done as you have? Do you covet the humble pittance which you have exchanged for this book? If so, then read your Bible, and when you find the lines that read thus, "he that giveth to the poor shall not lack," for it shall be restored unto you four-fold, but not if you have committed the deed covetously, for the Lord rewardeth the liberal giver; and certainly I have no reason to say that you are not a liberal person, for you have bought my book and paid for it manfully. But perhaps it would not be well to proceed further in this direction; had I a disposition to do so, I could build you up in your own estimation until you would feel as tall as the tower of Babel; and on the other hand I could make you feel as pentitent as Job did under his affliction. I could make your heart leap for joy, or your face drawn down as long as any criminal's that ever stood at the bar of our United States Court. Perhaps some of you may wonder why I write thus. I can explain myself in a very few words; it is because many persons in my situation are told in plain words that the county house is good enough for us. No matter in what society he ranked before the misfortune befell him, he is passed by his old associates, and if spoken to at all, it is *Good morning, Dick, or Harry,* or any other nickname; when in former times, with a graceful bow, he would be saluted as Mr. I do not refer to myself alone, for this is too much the case with many persons that have become afflicted in any direction, and their finances have diminished by paying an enormous price for a few doses of calomel or some other poisonous drug, with a slight portion of morphine to keep the patient quiet. Reader, I have been somewhat abused in this way, and, if I have been hasty in my remarks, forgive me, and I will inform you of the cause of my blindness. By my trade I am a blacksmith, and after leaving the water I went to work at my trade, and on the fourth day of June, 1855, I was employed in the shop of Mr. Henry Pollard, in the town of Charlotte, near the city of Rochester, in the state of New York, when a small piece of steel flew into my left eye. I immediately ran to doctor Jones' office, and the steel was immediately removed, but it had done its work; my eye was ruined, so that I could not distinguish light from darkness. The blow caused a great inflammation, which was exceedingly painful, and with the help of many skillful physicians I was unable to subdue the inflammation. My right eye then began to sympathize with the other. I then went to Cleveland, and was doctored under the charge of Professors Ackley and Garlick, and at the end of six months I lost my sight wholly and they pronounced me incurable, with the assurance that I should ever remain blind, which was a touching sentence; but, however hard it may be for a person to lose their sight, however lonesome a blind man's life may seem, how drearisome the time passes in the sunshine days of summer; when bright to you it is dark to me, when all things are of one color. And now to the purchasers of this book, my gratefulness and true respect,

I remain Yours,

J. C. M.

COLOPHON

The J. C. Mullett book, *A Five Years's Whaling Voyage,* was printed in Cleveland, Ohio, in 1859 by *Fairbanks, Benedict & Co., Printers.* The book is now mildly rare. The original edition was printed from handset type, as were virtually all books of the day or up until the 1880's when the first Linotypes came into use. This Ye Galleon edition was put together in the workshop of Glen Adams which is located in the pleasant country village of Fairfield, Washington. The pages were prepared by Gay Peterschick, Robin Pearson and Sharyn Brown. The text type was set on a Compugraphic 48 photosetter. The text type is Baskerville, running heads are 18 point Stephenson, Blake Bologna, page numbers are 24 point Washingon Text, the paper is Mustang in 100 pound weight. Binding is by William Bosch of Spokane. We had no special difficulty with the work. Pen and ink drawings are by Susan Hughes.

Books from the workshop of Glen Adams in Fairfield, Washington. Ye Galleon books carry a cargo of words. . . . as long, long ago in the sixteenth and seventeenth centuries the clumsy Spanish galleons carried rich cargoes from the New World to the Old.

Of this edition 423 copies were printed.

This is Copy Number 117.